SHARIE OROZCO

Dragonflies and Damselflies

First edition

This book was professionally typeset on Reedsy.
Find out more at reedsy.com

This book is dedicated to my daughters, Jaylee and Gwen, for supporting this crazy dream of mine. They didn't question it for a second, and actually pushed me to do it.

Contents

1

Introduction

I f you're reading this book, then you are probably as interested, or as obsessed with dragonflies and/or damselflies, as I am. Since I can remember, I have been fascinated with these beautiful creatures that can fly in what seems to be impossible ways. It's almost as if they are dancing an intricately choreographed recital in the air. To have a dragonfly land on me is still a hope, as I'm positive it brings good luck. And I could study their paper thin wings that glisten in the light for hours on end.

This fascination with dragonflies grew recently when a close friend of mine passed away. I have never experienced a heartache like that, and I was devastated that I would never see her again. And then I heard about a legend that angels come to visit us, riding on the wings of dragonflies. Watching dragonflies over the past few years is even more amazing to me, and it creates a more feelings for me. Seeing a dragonfly clinging to my screen door, perched on a flower in the garden, or just whooshing through the air past me brings a feeling of peace, and often times, a vision of my dear friend comes to mind.

These new feelings made me curious about dragonflies. I wanted to know more facts, and, since I believed in the legend that angels could ride on their wings, I wanted to know if there were other legends. I learned so many facts and legends that I felt the need to share them in a book. I hope you'll enjoy what I found, and maybe learn a little too.

2

Facts

Dragonflies and Damselflies

The first thing I should mention is that I learned that there are damselflies, as well as dragonflies, and they are common. If you are like me, you probably assumed that they were all dragonflies. I found that the one that I am actually the most fond of is the damselfly. They have longer and thinner bodies, a gap between their eyes, both forewings and hindwings are about the same size, and they usually fold their wings up above their body when they are at rest. In contrast, dragonflies have bulkier bodies that are shorter and thicker, their eyes take up most of their head, their hindwings are broader than their forewings, and their wings stay in a flat shape out to their sides when they are at rest.

Damselfy

Dragonfly

Species

There are approximately 5000 species of dragonflies known to humans. They are a part of the Odonata order of flying insects. Odonate means "toothed one" in Greek, referring to serrated teeth. They are considered carnivores as they eat other insects, and sometimes other dragonflies.

The most common species in the world is probably the Globe Skimmer Pantala Flavescens. It is cosmopolitan and can be found on all continents in the warmer region, while most Anisoptera species are tropical.

One of the rarest dragonflies in the world is the golden dragonfly and they are an endangered species. They are threatened by habitat loss and pesticides, and now there are only a few hundred in the world found in the tropical rainforests of South America and Africa. They are the largest dragonfly in the world with a wingspan of up to 8 inches.

Evolution

A man named Frank Carpenter discovered a fossil of a giant dragonfly (Meganeuropsis Permiana) in Elmo, Kansas in 1939. These prehistoric insects are known informally as "griffinflies." Experts say the wing-span measured around 2 ½ feet, and were estimated to weigh over 1 pound, which is similar to the weight of a crow. They dated the fossil back to the Permian era, which was about 300 million years old.

Another well preserved dragonfly fossil dating back about 200 million years ago was discovered in China in 2013. It is possible the prehistoric dragonfly was the largest insect that ever existed, and it is believed their large size was possibly due to high oxygen levels during that time. These dragonflies became extinct at the end of the Permian period, probably caused by many factors such as climate change, a drop in oxygen levels, and the arrival of the first birds that became predators.

Stages of Life

Dragonflies lay as many as 1500 eggs at a time, usually shaped like grains of rice, inside plant tissues. They can also be the size of a pinhead, and be ellipsoidal or nearly spherical. The eggs hatch in about a week into nymphs which molt 6 to 15 times as they grow, depending on the species. The nymph stage lasts up to 5 years for larger species, and between 3 months and 3 years for smaller species. They spend the majority of their life in this stage, beneath the water's surface.

Once the nymph is ready to metamorphose into its adult stage, it heads to the surface of the water, usually at night. It remains stationary with its head out of the water as it adapts to breathing air. It then climbs up a reed or other plant extending from the water and molt. The nymph exoskeleton will begin to split at a weak spot behind the head as it begins to crawl out while arching backwards until all but the tip of the abdomen is free, to allow the adult exoskeleton to harden. It then curls back upwards to complete the emergence, swallowing air to pump out its body, and pumping haemolymph into its wings so they can fully extend. The wings can take hours or days to harden. Most species have an adult life span of about 5 weeks or less, but they can survive up to 10 weeks.

Reproduction

When a male dragonfly finds a female he wants to mate with, he has to attract her to his territory while driving off rival males. When he feels that he is ready to reproduce, he transfers a packet of sperm from his primary genitalia on segment 9 of his abdomen, near the end, to his secondary genitalia on segments 2 and 3, near the base of his abdomen. Once he wins the female over, he grasps her by the head with his claspers at the end of his abdomen. They then fly in tandem or perch on a twig or stem with the male in front. The female will then curl her abdomen downwards and forward under her body to pick up the sperm packet. This position is commonly called the "heart" or "wheel."

The "heart"

Egg-laying then becomes a task with the female flying over floating or waterside vegetation to deposit the eggs while the male continues to grasp her head or while hovering over her to hopefully prevent other males from removing his sperm and inserting their own. This is possible due to delayed fertilization and sexual selection. If another male successfully takes over, he scrapes the previous inserted sperm out with his penis, and then inserts his own. When flying in tandem, the female glides which takes less effort and allows her more energy to expend on laying the eggs. Also, when the female submerges to deposit the eggs, the male can help pull her out of the water.

Habitat

Dragonflies species have a variety of habitats, but all are around water of some form. Some species prefer flowing waters, and some prefer standing water. Adult males will defend their territories near water to provide an environment for nymphs to develop, and for females to lay their eggs. This territorial behavior can be against other species of dragonflies, or against other predatory insects, such as flying ants or termites. Some display their ownership of an area with bright colors on their face, abdomen or wings. Others may have aerial dogfights, or high speed chases. Also of note, a female dragonfly must mate with a territory male before she can lay her eggs in that territory.

Flying

When it comes to flying, dragonflies are the specialists. They can fly in any direction and change direction suddenly. They are also very powerful fliers with the ability to migrate long distances, including across the sea. They have a very high power to weight ratio, and have been documented accelerating at nearly 4 G linearly, and 9 G in sharp turns. The most reliable speeds on record for some larger dragonflies are 22 - 24 mph. They can fly 100 body lengths per second going forward, and 3 lengths per second going backwards. Dragonflies have four different flying styles:

- Counter-stroke - The forewings will beat at 180 degrees out of sync of the hindwing. This allows it to hover and slow flight, and generates a large amount of lift.
- Phased-stroke - The hindwings beat 90 degrees ahead of the forewings. This is used for flight as it creates more thrust, but less lift.
- Synchronized-stroke - All 4 wings beat together. This is used when changing direction and creates maximized thrust.
- Gliding - The wings are held out. This style can be used for 3 situations: free gliding between bursts of flight, gliding in the updraft of the crest of a hill, and sometimes a female will glide

while mating with a male and he beats his wings.

Prey

In nymph state, dragonflies will eat almost anything that is smaller than them. This mostly consists of bloodworms and other insect larvae, but can also include tadpoles and small fish. A few species will leave the water to hunt small arthropods on the ground at night, and some even leap out of the water to attack full grown tree frogs.

As adults, dragonflies hunt while in flight using the incredibly acute vision and strong flight skills. They are mostly carnivorous and eat a wide variety of insects. This can include mosquitos, butterflies, moths, damselflies, and even smaller dragonflies. They will only eat what they catch when they are flying, and when caught, they will usually bite the head of its prey, and then carry it to a perch to consume. The wings are discarded, and then the head is usually ingested first. They are very efficient at hunting, catching up to 95% of the prey they pursue, and they can eat up to a fifth of their body weight in prey per day, which can be between 30 to hundreds of mosquitos.

Predators

Ducks and herons can be predators to nymphs in the water, in addition to newts, frogs, fish, and water spiders. Amur falcons also feed on globe skimmer dragonfly nymphs while on the wing over the Indian Ocean, as their migrations coincide with each other.

Even though adult dragonflies are swift and agile fliers, some predators are fast enough to catch them. Most of these predators are birds like falcons and swallows, although some species of wasps are known to prey on dragonflies as well. Wasps will use them as provision in their nests, laying an egg on each one they catch.

3

Legends and Meanings

During my research, I found that legends about dragonflies are plentiful. They vary from regions around the world, as well as across time periods. They can be heartwarming and pleasant, or ominous and even a little scary! While I prefer to stick with my belief of angels flying on the wings to come visit us from beyond, I did find these legends to be very interesting and fun to learn about.

Some legends I found were of good luck, or signs of the future, like some fishermen believe that if they see a dragonfly, the fishing will be good. Others believe that the color of dragonfly you encounter means something different will happen in the near future, or that it could be a sign for you to think about in your life.

Colors

It seems you can find dragonflies in almost any color imaginable. There are some colors that are more prominent, but finding one that is a striking color, or that's wings glisten in the light, is almost magical. There is the natural pigmentation, meaning the color that stays the same no matter how you look at it. These are usually brown, black, yellow, red or blue. Then there are structural colors, meaning the color that can appear different depending on how the light hits. These colors have a shimmering or iridescent effect, such as glistening wings. Here's a little bit about what some people believe each color of dragonfly means.

Blue - A very common color of dragonfly, and it can be found on the body, eyes, or wings. This color symbolizes loyalty, faith and trust. It is also the symbol of the fifth chakra (throat chakra) linked to how you connect and communicate with people.

If you see a blue dragonfly, it may be a sign to examine how you communicate with others and the messages you put out to the world. You can think of that as verbal, or it could also be facial expressions or body language. So you may want to consciously think about all your actions.

Examples of blue dragonflies are the blue emperor or blue dasher.

Blue Dragonfly

Green - This color can also be found on the body, eyes, or wings, and it is usually iridescent. It holds meanings of growth, fertility, abundance, and renewal. It is associated with the fourth chakra (heart chakra) linked to heart, lungs, and respiratory system. It is believed to affect serious emotions and relationships in life.

If you encounter a green dragonfly, it could mean you need to make some changes to relationships in your life. That could be relationships with a partner, friends, or family. Since green is the color of heart and nature, these dragonflies are believed to come by to encourage healthy, nurturing relationships.

Examples of green dragonflies are green darner, pondhawk, and the Giant Hawaiian.

Green Dragonfly

Red - This color is a natural pigmentation, usually in combination with other colors. It signifies strength, power, anger, and intensity, but it can be interpreted differently depending on the culture. Some see it as a sign of wealth and good luck, whereas others could see it as a sign of Autumn. Red is associated with the first chakra (root chakra) with meanings of power, passion, courage, and vitality. This chakra helps humans feel secure. Red dragonflies can be found all over the world, but are considered rare.

If a red dragonfly passes your way, it could be a sign that you need more passion or security. You could take as you need to find a way to become more grounded and determined in what you do.

An example of a red dragonfly is the red-veined darter.

15

Red Dragonfly

Yellow - It is rare to see an all yellow adult dragonfly, but some juvenile males are yellow. Other dragonflies have yellow in the form of stripes or paired with other colors in other ways. Yellow is a sign of happiness, honor, and optimism. It is associated with the third chakra (solar plexus) which is how you assert yourself in the world.

A yellow dragonfly could help you feel more youthful and exuberant. It may make you want you to accomplish more goals or make some positive changes in your life. If you are religious, you may take it as a sign to tap into your spiritual side.

Some examples of yellow dragonflies are the yellow-veined darter, black petaltail, and the river cruiser.

Yellow Dragonfly

Orange - This color signifies joy, wellness, and creativity. It is the color of the second chakra (sacral chakra) which is related to your gut instincts and sexuality.

An orange dragonfly could indicate that something in your life needs more nourishing. This could especially be related to your relationships and physical wellbeing. If your health and wellness is nourished, it could help you have happier emotions and to be more creative. It could also mean that you need to follow your gut. Stop wasting time avoiding decisions and just follow your instinct.

Some examples of orange dragonflies are the firecracker skimmer and flame skimmer.

Orange Dragonfly

Purple - this color shares some similar meanings with blue and red since it is a balance between the two colors. It can symbolize peace, nobility, aspiration, devotion and wealth. It is linked to the sixth chakra (third eye) which is related to mind and intuition, as well as the seventh chakra (crown chakra) which is related to knowledge and consciousness.

A purple dragonfly could inspire you to think deeper and trust your intuition, or encourage you to pray and meditate. For religious people, these dragonflies may be seen as a calling from a higher power.

Examples of purple dragonflies are the purple skimmer and the roseate skimmer.

Purple Dragonfly

Brown - Since brown dragonflies blend into earth surroundings such as wood, dirt and soil easier, this color is related to being grounded and down to Earth. Stability and security could also play a part with this color.

Some believe that brown dragonflies, as well as brown butterflies, are a sign that you need to ground yourself, focus on what is important in life, and work on building a foundation to reach your goals. You could also take them as a sign to reorganize your home to make them stress-free.

Examples of brown dragonflies are brown hawkers, swamp darners, and Norfolk hawkers.

Brown Dragonfly

Black - This color signifies mystery, elegance, and rebellion. Black is an essential color that adds hue and variance to all other colors, so these dragonflies are some of the most important you can come across. They are common, but often overlooked.

If a black dragonfly graces you with its presence, it could be to remind you that privacy and discretion are good. A lot of people seek attention and over-share nowadays, so it can be a nudge to keep to yourself more often. It could also mean that there may be an issue in your life that you need to take a look at.

Some examples of black dragonflies are black skimmers, black saddlebacks, and the giant petaltail.

Black Dragonfly

White - This is the color of purity and innocence. It is related to the seventh chakra which has to do with your mental and physical capabilities. White dragonflies are compared to angels because of purity and innocence. They are rare, so you'll usually see white paired with other colors.

You could take this color on a dragonfly as a sign from someone close who has passed on that any pain or discomfort they had is now gone, and it may help you may feel relief and peace. It could also remind you not to focus on pain, anguish, or even death. Focus on the good things in life and remember that you can miss those who have passed and still have a happy and fulfilling life.

Some examples of white dragonflies are the common whitetail and long-tailed skimmer.

White Dragonfly

Pink - This color is associated with love and kindness. It is rare to see it in nature, and pink dragonflies are not common, but when you come across it, it is gorgeous.

If you are lucky enough to find a pink dragonfly, you may take it as a sign that romance, and possibly love, are in your near future. Or, you could see it as a reminder to spread kindness to others. Either way, positivity is what you should take from it.

An example of a pink dragonfly is the roseate skimmer with a pink tail.

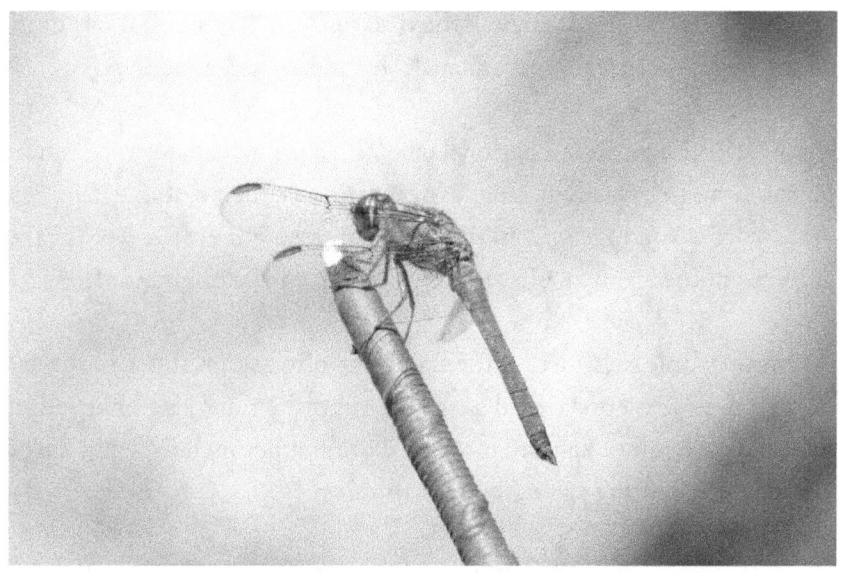

Pink Dragonfly

The next time you see a dragonfly, pay attention to the colors and think about what it means to you. I believe these creatures are special and can really help if you take a moment and evaluate your life. What positive changes can you make? How could it help those around you.

Legends

Beyond colors, as mentioned previously, there are many legends regarding dragonflies. Some are more detailed than others and can be very interesting.

Asia

Various countries in Asia view the dragonfly as a symbol of strength, courage, prosperity, success in battle, harmony, and good luck.

Japan - The dragonfly is a national emblem in Japan, and it was common for samurais to have dragonflies as decoration on their clothing, helmets, and swords. You may see children in Japan using their hands to make patterns in the air in an effort to mesmerize and catch dragonflies.

Japanese people believe that the red dragonfly carries the souls of the departed, so it is considered sacred. Dragonflies are also believed to bring good fortune. You can often find dragonflies mentioned in haiku poetry, representing strength and happiness.

You may hear Japan referred to as Akitsushima, or the "Island of the Dragonfly," because of the country's curved shape that looks a little like dragonflies mating, or a dragonfly at rest.

Map of Japan

Another legend is about the 21st emperor, Yuryaka Tenvo. While Yuryaka was out hunting, an insect bit his arm. Suddenly a dragonfly appeared, ate the insect, and rescued the emperor from further harm.

In August, Japan celebrates the Buddhist holiday of Bon or Obon. This is a celebration of their ancestors, and they believe the spirits of the deceased return for a visit. During this time there are thousands of dragonflies swarming the area, and it is thought that they are carrying the spirits of the departed. The people will build bonfires to guide the spirits home, and then build another fire to send them on their way again. During Bon, children are not allowed to catch dragonflies out of respect to the spirits. Families welcome the creatures into their home to temporarily rejoin their family

China - Many people in China believe that dragonflies bring good luck, prosperity, harmony, and help ward off evil spirits. They are featured in art and literature quite often. They also believe dragonflies have strong spiritual powers, and can be used in Feng Shui for positive energy flow in the home.

According to Chinese historical beliefs, the dragonfly is the insect that embodies the dragon's soul. Dragons are also loved and worshiped as can be seen in the many shrines and statues throughout the country. Dragons are believed to bring happiness, and the flying insect is considered a symbol of love.

Europe

European tales of dragonflies are not as joyful. Across Europe, dragonflies are often associated with horses, the devil, or black magic. Some of these tales do appear in American folklore, which indicates that people took the beliefs with them when they migrated to the New World. They have many nicknames including the devil's darning needle, and horse stinger.

It may be that the dragonfly is associated with the horse because it was commonly found around horses that were kicking and stomping. Although the dragonflies were eating the pests that were bothering the horses, such as mosquitos and flies, the people assumed that the dragonfly was the one doing the biting. Thus the nickname "horse stinger" came about.

In England, the people would tell their children that if they were bad, the dragonfly, or the devil's darning needle, could sew their month's shut while they sleep. Norwegian folklore has images of dragonflies poking holes in children's eardrums. Sweden calls dragonflies "blind stinger" that could poke your eyes out and "troll spindles" that were used for weaving as well as for poking out the eyes of your enemy.

Some of the names for dragonflies in Germany are Teufelsnadel ("Devil's needle"), Wasserhexe ("Water witch"), and Teufelspferd ("Devil's horse"). There are tales that the Germans believed dragonflies could sew your eyelids shut.

Germany's story of the dragonfly's origin is that a vicious princess rode around her kingdom without a care. One day she came across a small man who tried to speak to her. She told him to get out of her way, but he didn't listen, so she rode over him. The man then put a curse on her to always be joined to her horse. She and her horse became one, and transformed into the dragonfly.

Romanians have similar beliefs to Germany. Some people believe dragonflies came from Romania, and some believe they came from Germany.

There is a tale from Romania that the devil tried to cross a lake. He asked a fisherman for a ride in his boat, but the fisherman refused his request. The devil then adopted the form of the dragonfly, and flew across the lake. This is how some believe dragonflies came about.

The Celtic folklore associates dragonflies with snakes. They believe dragonflies protect poisonous snakes, or "Adder's Servant." Dragonflies were sometimes spotted sitting on the heads of snakes, as if they had charmed them into a trance. They also believed that dragonflies would protect the lives of injured snakes by stitching their wounds.

America

In America, Native Americans believe that dragonflies were originally dragons. One of the tales is that a coyote tricked a dragon into shape shifting, but then it couldn't change back. This brought belief that dragonflies symbolize change, speed and illusion.

You can find dragonfly symbols in Native American art, jewelry, and pottery. The images may show them near water, representing the water's purity. Their presence in nature was linked to the success of their corn crops. They could also be a sign that it was going to rain. If they were flying high, it meant heavy rain. If they were flying low, it meant a lighter rain.

A few more random superstitions I came across are:

- If a dragonfly lands on you, it is good luck, or you'll hear good news from someone you care about.
- Seeing a dead dragonfly means you'll hear some sad news.
- Catching a dragonfly means you'll marry within the year.

Regardless of what you believe a dragonfly means, I think we can all agree these creatures play a very important role in the world. If for nothing else, to eat those pesky mosquitoes!

4

Science

Scientists are intrigued with these amazing insects because of their flying skills. They are trying to mimic their flying ability and have done research for biomimetic applications. Dragonfly wings are mainly composed of veins and membranes that have a complex design that give rise to whole-wing characteristics, which makes them extremely versatile and maneuverable fliers. The wing structure is believed to be the reason for their enhanced aerodynamics. Scientists are trying to understand the mechanical properties of the wings so they can perform simulated models.

5

Conclusion

I thoroughly enjoyed writing this book, and learning so much about dragonfly facts and legends. I hope this book gave you a broader glimpse into them, and maybe even helped you appreciate them a little more. Maybe next time you see a dragonfly you'll stop to study it a little longer...if it will let you. Or if you are lucky and you have one land on you, perhaps you'll think of a loved one who passed on and tell them hello, that you miss them, but also acknowledge that they are there to tell you they are happy in their new place. There is so much beauty all around us, and these amazing dragonflies help enhance that beauty.

If you enjoyed this book, I would truly appreciate it if you'd leave a review for it on Amazon!